Dimensional Analysis
&
Conversion Factors

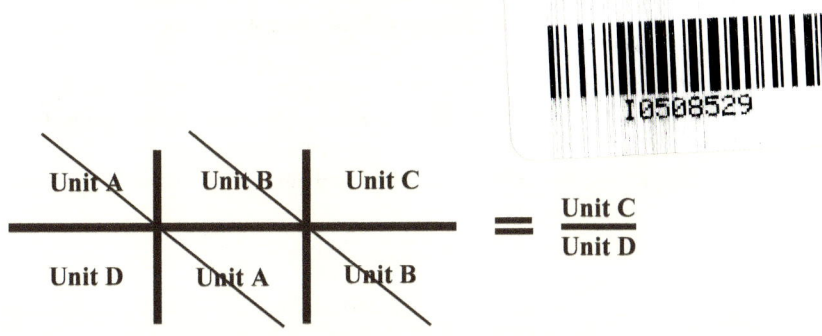

Merle A. Barlow

WESTBOW
PRESS®
A DIVISION OF THOMAS NELSON
& ZONDERVAN

Copyright © 2018 Merle A. Barlow.

All rights reserved. No part of this book may be used or reproduced by any means, graphic, electronic, or mechanical, including photocopying, recording, taping or by any information storage retrieval system without the written permission of the author except in the case of brief quotations embodied in critical articles and reviews.

WestBow Press books may be ordered through booksellers or by contacting:

WestBow Press
A Division of Thomas Nelson & Zondervan
1663 Liberty Drive
Bloomington, IN 47403
www.westbowpress.com
1 (866) 928-1240

Because of the dynamic nature of the Internet, any web addresses or links contained in this book may have changed since publication and may no longer be valid. The views expressed in this work are solely those of the author and do not necessarily reflect the views of the publisher, and the publisher hereby disclaims any responsibility for them.

Any people depicted in stock imagery provided by Getty Images are models, and such images are being used for illustrative purposes only. Certain stock imagery © Getty Images.

ISBN: 978-1-9736-2061-7 (sc)
ISBN: 978-1-9736-2060-0 (hc)
ISBN: 978-1-9736-2062-4 (e)

Library of Congress Control Number: 2018902267

Print information available on the last page.

WestBow Press rev. date: 03/13/2018

This book is dedicated to those persons who would perform a calculation involving a series of multiplications and divisions, without giving any consideration to the units of measurements associated with the numbers. Consequently, they may discover that the resulting answer is <u>incorrect</u>. These persons must learn that the multiplication and division of the units of measurement must also be correct in order for the numerical answer to be correct.

Preface

Cancellation Process for Units of Measurement

An important consideration to remember when performing calculations involving units of measurement is that if you want to eliminate a unit of measurement that is in the numerator of a ratio, make sure that the same unit is in the denominator of some ratio. The opposite is also true. If some unit that you want to eliminate is in the denominator of a ratio, make sure that the same unit is also in the numerator of some ratio.

Dimensional Analysis is essentially the conversion from one unit of measurement to another unit.

Clearwater, Florida
February, 2018

Merle

Prologue

For the purpose of Factor Conversion and Dimensional Analysis, let us examine the specific details of the "Unit Analysis Bridge" that is on the cover of this book, and is also presented below. This process of analysis is used throughout all the examples in this book.

Unit A	Unit B	Unit C
Unit D	Unit A	Unit B

The top horizontal row of numbers are involved in multiplication.
The bottom horizontal row of number are involved in multiplication.
The numbers in the vertical columns are involved in division.

The numbers and units of measurement in the top row represent the numerators of ratios.
The numbers and units of measurement in the bottom row represent the denominators of ratios.

ANY DIMENSIONAL UNIT in any numerator can be cancelled with ANY EQUAL DIMENSIONAL UNIT in any denominator.

The horizontal lines and the vertical lines representing the "Analysis Bridge" is presented here for the purpose of explaining the cancellation actions in the dimensional analysis process. **These actual lines will not appear in the example problems presented, but they can be easily imagined.**

CONTENTS

Title Page i

Dedication iii

Preface v

Prologue vii

What is Dimensional Analysis? xiii

Introduction to Dimensional Analysis xv

Note to Reader xvi

Section 1

1 $S = \frac{1}{2} at^2$ 2

2 $F = Ma$ 4

3 $E = Mc^2$ 6

4 $D = rt$ 8

5 Rectangular Solid (Parallelepiped) 10

6 Sphere 13

7 Cylinder 16

8 Cone 19

9 Pyramid 22

10 Cube 25

11 Simple Interest 28

12 Compound Interest 30

13 Continuous Compounding Interest 32

Section 2

1 Meters to Centimeters 35

2 Centimeters to Meters 35

3 Centimeters to Millimeters 35

4 Grams to Kilograms 35

5 Meters to Kilometers 36

6 Cubic Centimeters to Cubic Meters 36

7 Furlongs per Fortnight to Miles per Hour 36

8 Yards per Second to Miles per Hour 37

9 100 Yards compared to 100 Meters 37

10 One Yard to Meters 38

11 Number of Seconds in 7 Days 38

12 One Kilometer of Pennies 38

Section 3

1 A 10K Race 40

2 Two Perspectives of Car Velocity 40

3 World Water-ski Championship 40

4 Miles per Hour compared to Feet per Second 41

5 Volume Comparison 41

6 Dimensional and Numerical Validity 42

7 Kilometers per Hour compared to Meters per Second 43

8 Astronomical Distances 44

9 Two Perspectives of a 24-Hour Day 44

Section 4

1 Cubical Water Tank 47

2 A Million Dollars 48

3 A Billion Dollars Around the World 49

4 Chess – A Plethora of Possibilities 50

5 Dimensions of Heaven 51

6 Counting to a Million 52

7 Two Painters 53

8 Hens & Eggs 54

9 World Record 56

10 100 Meters compared to 100 Yards (Redux) 57

11 Dimensional Analysis Challenge – The Bondoraxpas 58

12 A Mathematical Challenge 59

Section 5

A Potpourri of Thoughts and Ideas

Epilogue 97

Postface 98

Appendix 99

What is Dimensional Analysis?

The dimensions of a measurement refer to the type of units for a quantity. For example, length is a dimensional quantity that can be measured in meters, centimeters, kilometers, miles, yards, feet, inches, etc. Dimensional analysis is the process of solving algebraic equations for units as well as numbers. It is a way of checking to ensure that you have used the correct equation, and that you have correctly applied the rules of algebra when solving the equation. It is a good practice to make dimensional analysis a habit by always stating the units as well as the numerical values whenever substituting values into an equation. It is necessary to have equivalent dimensions on both sides of the equation. Dimensional analysis is the process of carrying units of measurement throughout a calculation.

If we want to know the speed of a runner, a ball, a car, a missile, or anything else that moves, we measure the distance between two recognizable points. We measure the time it takes the runner, for example, to traverse that distance. We divide the distance by the time, and the result is the *speed* of the runner, or of any moving object. The answer is perhaps 10 feet per second, 30 meters per second, 1000 kilometers per hour, or about 186,000 miles per second (the speed of light). All these examples are *speeds*.

The concept of speed is defined as a distance divided by time. The meaning of speed is given by the formula:

$$\text{speed} = \frac{\text{distance}}{\text{time}}$$

In dimensional theory, instead of the word *distance*, we write, by convention, the symbol L (for length). Instead of the word *time*, we write T. Then we say the *dimension* of speed (speed in general) is L over T.

What does it mean to divide 1000 feet by 30 seconds? What does "32 feet per second per second" mean? One learns that 1000 feet over 30 seconds is a speed of $33\frac{1}{3}$ ft/sec, and that 32 feet per second per second is "g", the acceleration due to gravity. Slowly, these words begin to make sense. Speed makes sense, then acceleration. Mass comes with more difficulty. When these ideas make sense to you, you are ready for dimensional theory. Then you will understand that L/T is for speed, L/T^2 is for acceleration, and ML/T^2 is for force, because force is mass times acceleration.

Introduction to Dimensional Analysis

In the study of Mechanics within the discipline of Physics, the fundamental dimensions are:

M, L, T

Some of the example problems in this book use these three dimensions. However, most of the examples use only the L and T dimensions.

Mass   M   kg, gm, lb

Distance   L   m, cm, ft, mi

Time   T   sec, min, yr

Any quantity, such as distance, height, length, width, etc., that is or could be measured in terms of the meter, feet, etc., is said to have the DIMENSION of *length*. Similar definitions apply to the dimensions of *mass* and *time*.

The capital letters M, L, and T are dimensional symbols used as a shorthand for the dimensions of mass, length, and time.

Dimensional symbols are enclosed in <u>square brackets</u> to indicate that they express only the *dimension* of a quantity, and <u>not</u> its *numerical* value.

If a quantity has the dimension of length, this is written in dimensional notation as: $[L]$.
If a quantity has the dimension of time, this is written in dimensional notation as: $[T]$.
If a quantity has the dimension of mass, this is written in dimensional notation as: $[M]$.
For example:
In the equation $F = Ma$, $[a] = \left[\dfrac{L}{T^2}\right]$, and

$[F] = [MLT^{-2}] = \left[\dfrac{ML}{T^2}\right]$.

Note to Reader:
The presentation of the many examples in the following sections will illustrate:

(1) The conversion from one dimensional unit of measurement to another dimensional unit of measurement.

(2) The units of measurement, as well as the numerical values, must also be processed correctly mathematically in order to obtain a correct result.

(3) The dimensions on both sides of an equation must be equal.

The various examples presented will vary in detail, explanation, or elaboration. This is by intent, so that there is not an overwhelming preponderance of complexity.

Section 1

All examples in this section illustrate the solution of equations for the Units of Measurement only. **There are no numerical calculations.**

1

$$S = \frac{1}{2}at^2$$

This is an equation that determines the distance traveled by an object starting at rest.

1

$$S = \frac{1}{2}at^2$$

The variables of this equation are:

S
The variable (S) is the distance traveled in one of several optional lengths. In this example, the option will be meters (m).

a
The variable (a) is the acceleration of the object, measured in miles per second squared ($m/\sec^2$).

t
The variable (t) is the time required in seconds squared ($\sec^2$).

$$S = \frac{1}{2}at^2$$

Following is a Dimensional Analysis (D/A) evaluation of this equation:

[distance] = [L]

$[a] = [LT^{-2}]$

$[t^2] = [T^2]$

$[L] = \frac{1}{2}[LT^{-2}(T^2)]$

$[L] = [L]$

Note that the dimensions on both sides of the equation are equal, and therefore the equation is <u>dimensionally</u> correct.

2

$$F = Ma$$

> The standard unit of force in the International System of Units (SI) is the Newton (N). A Newton is equal to the force that produces an acceleration of one meter per second per second on a mass of one kilogram.

2

$$F = Ma$$

The variables of this equation are:

F
The variable (F) is the force that produces an acceleration of one meter per second per second is equal to 1N (Newton).

M
The variable (M) is the mass of one kilogram that is accelerated one meter per second per second that produces a force of 1 Newton.

a
The variable (a) is the acceleration of one meter (m) per second per second $\left[\dfrac{L}{T^2}\right]$.

The definition of force is:

$$(1\,kg)\left(\dfrac{1m}{\sec^2}\right) = 1N \text{ (Newton)}.$$

Therefore, the Unit of Measurement for Force is:

$$Force = (1\,kg)\left(\dfrac{1m}{\sec^2}\right)$$

$$[Force] = [M]\left[\dfrac{L}{T^2}\right]$$

$$[Force] = \left[\dfrac{ML}{T^2}\right]$$

3

$$E = Mc^2$$

> This equation expresses the equivalence of Energy and Mass, and the fact that either can be transformed into the other. Energy is the product of a Mass and the speed of light squared.

3

$$E = Mc^2$$

The variables of this equation are:

E
The variable (E) represents the Energy (the equal relationship between energy and mass) measured in Joules.

A Joule is defined as the force that produces an acceleration of one meter per second per second on a mass of one kilogram. This force is known as a Newton.

M
The variable (M) represents the mass of an object, and is measured in kilograms.

c
The variable (c) representing the speed of light is approximately 300,000,000 meters per second.

$F = Ma$

$$[Force] = \left[\frac{ML}{T^2}\right]$$

$(F)(L) = Joule$

$(F)(L) = ML^2T^{-2}$

$c^2 = (LT^{-1})^2 = L^2T^{-2}$

$E = Mc^2$

$[E] = \left[\dfrac{ML^2}{T^2}\right]$ (This is the dimension of Energy).

4

$$D = rt$$

> This equation establishes the relationship that a distance traveled is the product of the rate of travel and the time.

4

$$D = rt$$

The variables of this equation are:

D
The variable (D) represents a distance traveled at a specific rate and time.

r
The variable rate (r) represents a specific rate of speed.

t
The variable time (t) represents a specific time of travel.

$$D = rt$$

$$[L] = \left[\frac{L}{T}(T)\right]$$

$$[L] = [L]$$

A dimensional analysis of the individual variables:

$$r = \frac{L}{T}$$

$$t = [T]$$

$$rt = \left[\frac{L}{T}(T)\right]$$

Therefore, $[L] = [L]$.

The D/A for the Distance is $[L]$.

5

Surface Area & Volume of a Rectangular Solid

$$A = 2lw + 2hw + 2lh$$
$$V = lwh$$

> The **Surface Area** of a Rectangular Solid is equal to two times the product of the length and width, plus two times the product of the height and width, plus two times the product of the length and height.
>
> The **Volume** of a Rectangular Solid is equal to the product of the length, width, and height.

5

Surface Area
$$A = 2lw + 2hw + 2lh$$

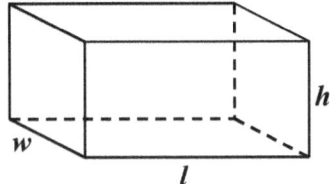

The dimension of Area is $[L^2]$.

Each of the three dimensions $= [L]$.

$l = [L]$; $w = [L]$; $h = [L]$

$lw = [L^2]$; $hw = [L^2]$; $lh = [L^2]$

$A = 2lw + 2hw + 2lh$

$[L^2] = 2[L^2] + 2[L^2] + 2[L^2]$

$[L^2] = [L^2]$

The D/A for the Surface Area is $[L^2]$.

5

Volume
$$V = lwh$$

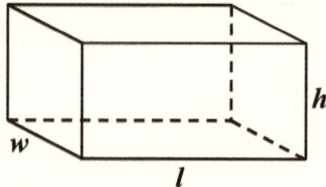

The dimension of Volume is $[L^3]$.

Each of the three dimensions = $[L]$.

$l = [L]$; $w = [L]$; $h = [L]$

$V = lwh$

$[L^3] = [L][L][L]$

$[L^3] = [L^3]$

The D/A for the Volume is $[L^3]$.

6

Surface Area & Volume of a Sphere

$$A = 4\pi r^2$$

$$V = \frac{4}{3}\pi r^3$$

The **Surface Area** of a Sphere is equal to four times the product of pi and the radius squared.
The **Volume** of a Sphere is equal to four-thirds times the product of pi and the radius cubed.

6

Surface Area of a Sphere

$$A = 4\pi r^2$$

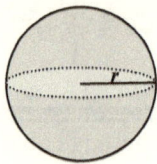

The dimension of Area is $[L^2]$.

The numbers "4" and "π" are dimensionless numbers.

r is a length = $[L]$.

r^2 is a length times a length = $[L^2]$.

The product of two dimensionless numbers and $r^2 = [L^2]$.

Therefore, in $A = 4\pi r^2$,
$$[L^2] = [L^2].$$

The D/A for the Surface Area is $[L^2]$.

6

Volume of a Sphere

$$V = \frac{4}{3}\pi r^3$$

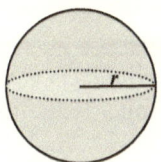

The dimension of Volume is $[L^3]$.

The numbers "$\frac{4}{3}$" and "π" are dimensionless numbers.

r is a length = $[L]$.

r^3 is a length times a length times a length = $[L^3]$.

The product of two dimensionless numbers and r^3 = $[L^3]$.

Therefore, in $V = \frac{4}{3}\pi r^3$,

$$[L^3] = [L^3].$$

The D/A for the Volume is $[L^3]$.

7

Surface Area & Volume of a Cylinder

$$A = 2\pi r^2 + 2\pi rh$$
$$V = \pi r^2 h$$

> The **Surface Area** of a Cylinder is equal to twice the product of pi and the radius squared plus twice the product of pi times the radius times the height.
>
> The **Volume** of a Cylinder is equal to the product of the base area (πr^2) and the height.

7

Surface Area of a Cylinder
$$A = 2\pi r^2 + 2\pi rh$$

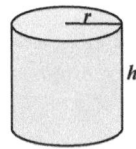

The dimension of Area is $[L^2]$.

The numbers "2" and "π" are dimensionless numbers.

r is a length = $[L]$.

r^2 is a length times a length = $[L^2]$.

The product of two dimensionless numbers and $r^2 = [L^2]$.

The product of two dimensionless numbers times the product of r and $h = [L][L] = [L^2]$.

The sum of $[L^2]$ and $[L^2] = [L^2]$.

Therefore, in $A = 2\pi r^2 + 2\pi rh$,
$$[L^2] = [L^2].$$

The D/A for the Surface Area of a Cylinder is $[L^2]$.

7

Volume of a Cylinder
$$V = \pi r^2 h$$

The dimension of Volume is $[L^3]$.

The number "π" is a dimensionless number.

r is a length = $[L]$.

r^2 is a length times a length = $[L^2]$.

h is a length = $[L]$.

The product of a dimensionless number times the product of r^2 and h = $[L^2][L]$ = $[L^3]$.

Therefore, in $V = \pi r^2 h$,
$$[L^3] = [L^3].$$

The D/A for the Volume of a Cylinder is $[L^3]$.

8

Surface Area & Volume of a Cone

$A = \pi r l + \pi r^2$, where l = slant height

$$V = \frac{1}{3}Bh, \text{ or } V = \frac{1}{3}\pi r^2 h$$

The **Surface Area** of a Cone is equal to (pi times the radius times the slant height) plus the Base area.

The **Volume** of a Cone is equal to one-third times (the product of the Base area and the height).

8

Surface Area of a Cone

$A = \pi r l + \pi r^2$, where l = slant height

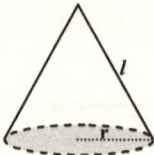

The dimension of Area is [L²].

The number "π" is a dimensionless number.

r is a length = [L].

r² is a length times a length = [L²].

l, the slant height of the Cone is a length = [L].

Therefore, in $A = \pi r l + \pi r^2$,
$$[L^2] = [L^2] + [L^2]$$
$$[L^2] = [L^2]$$

The D/A for the Surface Area of a Cone is [L²].

8

Volume of a Cone

$$V = \frac{1}{3}\pi r^2 h$$

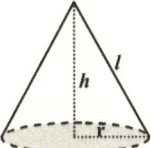

The dimension of Volume is $[L^3]$.

The numbers "$\frac{1}{3}$" and "π" are dimensionless numbers.

r is a length = $[L]$.

r^2 is a length times a length = $[L^2]$.

h is a length = $[L]$.

Therefore, in $V = \frac{1}{3}\pi r^2 h$,

$$[L^3] = [L^2][L]$$
$$[L^3] = [L^3]$$

The D/A for the Volume of a Cone is $[L^3]$.

9

Surface Area & Volume of a Pyramid

$$A = \frac{1}{2}Pl + B$$

$$V = \frac{1}{3}Bh$$

> The **Surface Area** of a Pyramid is equal to one-half times (the product of the Base Perimeter and the Slant Height) plus the Base area.
> The **Volume** of a Pyramid is equal to one-third times (the product of the Base area and the height).

9

Surface Area of a Pyramid

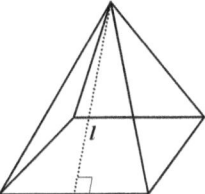

The dimension of Area is [L²].

The number "$\frac{1}{2}$" is a dimensionless number.

P, the Perimeter of the Pyramid Base, is a length = [L].

l, the Slant Height of the Pyramid, is a length = [L].

B, the Base Area of the Pyramid, = [L²].

Therefore, in $A = \frac{1}{2} Pl + B$,

$$[L^2] = [L][L] + [L^2]$$

$$[L^2] = [L^2] + [L^2]$$

$$[L^2] = [L^2]$$

The D/A for the Surface Area is [L²].

9

Volume of a Pyramid

$$V = \frac{1}{3}Bh$$

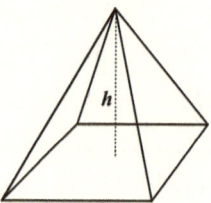

The dimension of Volume is [L³].

The number "$\frac{1}{3}$" is a dimensionless number.

B, the Base Area of the Pyramid, is a length = [L²].

h, the Height of the Pyramid, is a length = [L].

Therefore, in $V = \frac{1}{3}Bh$,

$$[L^3] = [L^2][L]$$

$$[L^3] = [L^3]$$

The D/A for the Volume is [L³].

10

Surface Area & Volume of a Cube

$$A = 6e^2$$
$$V = e^3$$

> The **Surface Area** of a Cube is equal to six times the square of an edge.
> The **Volume** of a Cube is equal to the cube of an edge.

10

Surface Area of a Cube
$$A = 6e^2$$

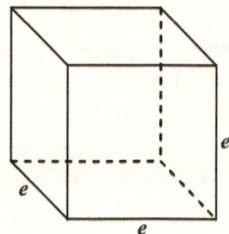

The dimension of Area is $[L^2]$.

The number "6" is a dimensionless number.

e, one of the twelve equal edges of a cube, is a length = $[L]$.

e^2 is a length times a length = $[L^2]$.

Therefore, in $A = 6e^2$,
$$[L^2] = [L][L]$$
$$[L^2] = [L^2]$$

The D/A for the Surface Area of a Cube is $[L^2]$.

10

Volume of a Cube

$$V = e^3$$

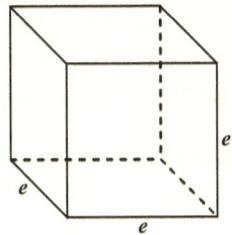

The dimension of Volume is [L³].

e, one of the twelve equal edges of a cube, is a length = [L].

e^3 is a length times a length times a length = [L³].

Therefore, in $V = e^3$,
$$[L^3] = [L][L][L]$$
$$[L^3] = [L^3]$$

The D/A for the Volume of a Cube is [L³].

11

Simple Interest Formula
$$I = \Pr t$$

> This formula represents the amount of Interest (I) earned when a Principal amount (P) is invested at an annual rate of interest (r), for a period of (t) years. The Interest earned is the product of the Principal, the rate, and the time.

11

Simple Interest Formula
$$I = \Pr t$$

The dimensional unit for I is [$].

The dimensional unit for P is [$].

r is the annual interest rate, or per cent per year $\left(\dfrac{\%}{yr}\right)$.

t is the number of years.

Therefore, the product, $rt = \left(\dfrac{\%}{\cancel{yr}}\right)(\cancel{yr}) =$ a per cent.

A per cent is a dimensionless number.

Therefore, in $I = \Pr t$.
$$[\$] = [\$].$$

The D/A for I is [$].

12

Compound Interest Formula

$$A = P\left(1 + \frac{r}{n}\right)^{nt}$$

This formula represents the amount of Interest (A) accrued when a Principal amount (P) is invested at an annual rate (r), and the interest is computed (n) times per year, and (t) is the number of years. Compound interest is interest computed on the investment as well as any accumulated interest.

12

Compound Interest Formula

$$A = P\left(1+\frac{r}{n}\right)^{nt}$$

The dimensional unit for A is [$].

The dimensional unit for P is [$].

r, n, and t are dimensionless numbers.

$\frac{r}{n}$ is dimensionless, because a fraction is a dimensionless number.

nt is dimensionless, because the product of two dimensionless numbers is a dimensionless number.

$\left(1+\frac{r}{n}\right)^{nt}$ is a dimensionless number raised to a power of a dimensionless number.

Therefore, $\left(1+\frac{r}{n}\right)^{nt}$ is a dimensionless number.

Hence, in $A = P\left(1+\frac{r}{n}\right)^{nt}$,

$$[\$] = [\$]$$

The D/A for A is [$].

13

Continuous Compounding Formula
$$A = Pe^{rt}$$

> This formula represents the amount of Interest (A) accrued when a Principal amount (P) is invested at an annual rate r for a period of (t) years, and the interest is compounded continuously. Although continuous compounding sounds good, it yields only a fraction of a percent more interest over a year than the interest from daily compounding.

13

Continuous Compounding Formula
$$A = Pe^{rt}$$

The dimensional unit for A is [$].

The dimensional unit for P is [$].

e is the base of the natural system of logarithms, and is equal approximately to 2.71828182..., and is a dimensionless number.

r and t are dimensionless numbers.

In e^{rt}, a dimensionless number, (e), is raised to the product of two dimensionless numbers, and the result is a dimensionless number.

Therefore, in $A = Pe^{rt}$,
$$[\$] = [\$]$$

The D/A for A is [$].

Section 2

The examples in this section illustrate Conversion Factors, the cancellation process for units of measurement, and the numerical result for the example problems. The solution for each problem is a complete Dimensional Analysis solution.

(1)
Convert 4m to cm.

Because there are 100 centimeters in one meter, the conversion of 4 meters to centimeters is as follows:

$$(4\,\cancel{m})\left(\frac{100\,cm}{1\,\cancel{m}}\right) = 400\,cm.$$

In the conversion equation, the meter dimension cancels to yield the result in centimeters.

(2)
Convert 50 cm to m.

Because there are 100 centimeters in one meter, the conversion of 50 centimeters to meters is as follows:

$$(50\,\cancel{cm})\left(\frac{1\,m}{100\,\cancel{cm}}\right) = \frac{50}{100}\,m = \frac{1}{2}\,m.$$

In the conversion equation, the centimeter dimension cancels to yield the result in meters.

(3)
Convert 15 cm to mm.

Because there are 10 mm in one cm, the conversion from 15 centimeters to millimeters is as follows:

$$(15\,\cancel{cm})\left(\frac{10\,mm}{1\,\cancel{cm}}\right) = 150\,mm.$$

(4)
Convert 8.3×10^4 gm to kg.

There are 10^3 gm = 1000 gm in 1 kilogram.

Therefore, $(8.3 \times 10^4\,\cancel{gm})\left(\dfrac{1\,kg}{10^3\,\cancel{gm}}\right) = 8.3 \times 10\,kg = 83\,kg.$

(5)
Convert 8.4×10^{10} m to km.

There are 10^3 m = 1000m in 1 kilometer.

$$\left(8.4 \times 10^{10} \cancel{m}\right)\left(\frac{1 km}{10^3 \cancel{m}}\right) = 8.4 \times 10^7 \text{ km} = 84{,}000{,}000 \text{ km}.$$

(6)
Convert 9.4×10^{12} cm^3 to m^3.

There are 10^6 cm^3 = 1,000,000cm^3 = 1m^3.

$$\left(9.4 \times 10^{12} \cancel{cm^3}\right)\left(\frac{1 m^3}{10^6 \cancel{cm^3}}\right) = 9.4\text{m} \times 10^6\text{m}^3 = 9{,}400{,}000\text{m}^3.$$

(7)
In this problem, we will direct our attention to the dimensional analysis regarding the conversion of furlongs per fortnight to miles per hour.

A furlong is exactly one-eighth of a mile and is a unit of distance that is common to horse racing. A fortnight is exactly two weeks, or 336 hours.

Convert $\dfrac{13{,}440 \text{ furlongs}}{\text{fortnight}}$ to $\dfrac{\text{miles}}{\text{hour}}$.

$$\left(\frac{13{,}440 \text{ furlongs}}{\text{fortnight}}\right)\left(\frac{\frac{1}{8}\text{mi}}{\text{furlong}}\right)\left(\frac{1 \text{ fortnight}}{336 \text{ hours}}\right)$$

$$(13{,}440)\left(\frac{1}{8}\text{mi}\right)\left(\frac{1}{336 \text{ hr}}\right) = \frac{13{,}440 \text{ mi}}{(8)(336)\text{hr}} = \frac{5 \text{ mi}}{\text{hr}}$$

13,440 furlongs per fortnight converts to 5 miles per hour.

(8)
A runner runs 100 yards in 10 seconds.
What is the runner's rate in miles per hour?

$$\left(\frac{100\,yd}{10\,sec}\right)\left(\frac{3600\,sec}{1\,hr}\right)\left(\frac{1\,mi}{1760\,yd}\right) = \frac{360{,}000\,mi}{17{,}600\,hr} = \frac{20.45...mi}{hr}$$

So, running 100 yards in 10 seconds is equivalent to running 20.45...miles per hour.

(9)
If a runner runs 100 yards in 10 seconds, what time is required to run 100 meters?
(Assume the same rate is maintained).

1 meter = 39.37 inches (exactly).

$$\frac{20.45\,mi}{hr} = 100\,yd\,rate.$$

100 meters is exactly 337 inches longer than 100 yards.
So, running at the 100 yard rate, how long will be required to run the additional 337 inches?

$$\frac{337\,in}{63{,}360\,\frac{in}{mi}} = .005319\,mi.$$

So, 337 in = .005319 mi.

$$\frac{.005319\,mi}{20.45\,\frac{mi}{hr}} = .00026\,hr.$$

This means that running .005319 mi at the rate of 20.45 miles per hour requires a time of approximately .00026 hr.

$$(.00026\,hr)\left(3{,}600\,\frac{sec}{hr}\right) = .936\,sec.$$

Therefore, to run a distance of 100 meters at the rate of 20.45 miles per hour will require a time of:
10 sec + .936 sec = 10.94 sec.

So, if you run 100 yards in 10 seconds, and maintain the same rate for 100 meters, the time required will be 10.94 sec to run 100 meters.

(10)
Convert one yard to meters.

$$\left(\frac{36\,in}{1\,yd}\right)\left(\frac{1\,m}{39.37\,in}\right) = .9144\,\frac{m}{yd}.$$

Therefore, one yard is less than one meter.
One yard is 91.44% of one meter.

(11)
How many seconds are in 7 days?

$$(7\,days)\left(\frac{24\,hr}{1\,day}\right)\left(\frac{60\,min}{hr}\right)\left(\frac{60\,sec}{1\,min}\right) = (7)(24)(60)(60)\,sec$$

7 days = 604,800 sec.

(12)
How many pennies, each having a diameter of about 1.9cm, positioned side by side in a straight line, would be required to extend to a length of one-thousand meters (1km)?

Number of Pennies =

$$\left(\frac{1\,km}{1.9\,cm}\right)\left(\frac{100\,cm}{1\,m}\right)\left(\frac{1000\,m}{1\,km}\right) = \frac{(100)(1000)}{1.9} = \frac{10^5}{1.9} = 52,632$$

Note that all units of measurement in both numerators and denominators cancel to yield the number of pennies required to form a 1km length of pennies.

Section 3

The examples in this section continue to illustrate conversion factors, the cancellation process for units of measurement, and the complete dimensional analysis solution for each problem (the D/A and numerical values).

(1)
A 10K race is 10 kilometers (km) in length. Using Dimensional Analysis, find the length of the race in miles (mi).
[1 meter (m) = 1.094 yards (yd)].

$$(10\,km)\left(\frac{1000\,m}{1\,km}\right)\left(\frac{1.094\,yd}{1\,m}\right)\left(\frac{1\,mi}{1760\,yd}\right) = \frac{10{,}940\,mi}{1760} = 6.2\,mi$$

The distance is approximately 6.2 miles.
The following units of measurement appear both as numerators and denominators, and therefore <u>cancel each other</u>:
km; m; yd
The unit of measurement remaining is miles as a numerator. This was the desired result of converting kilometers to miles.

(2)
A car travels a distance of 100ft in about 2.8sec. What is the average velocity of the car in miles per hour? Express the answer to the nearest whole number.

$$\left(\frac{100\,ft}{2.8\,sec}\right)\left(\frac{3600\,sec}{1\,hr}\right)\left(\frac{1\,mi}{5280\,ft}\right) = \frac{24\,mi}{hr}$$

The answer is approximately 24 miles per hour.
Note that both ft and sec cancel because they both are in numerators and denominators.
That leaves mi as a numerator and hr as a denominator, so the result is expressed as miles per hour.

(3)
A person won a World Water ski Championship race by finishing an 88 kilometer (km) race in 51.23 minutes (min). What was the average speed in miles per hour?
[1 km = 0.62 mi].

$$\left(\frac{88\,km}{51.23\,min}\right)\left(\frac{0.62\,mi}{1\,km}\right)\left(\frac{60\,min}{1\,hr}\right) = \frac{3273.6\,mi}{51.23\,hr} = \frac{63.9\,mi}{hr}$$

The average speed is approximately 63.9 miles per hour.

Note that in the calculation, km and min cancel. That leaves mi in the numerator and hr in the denominator, so that the answer is expressed as miles per hour.

(4)
Convert the speed of 63.9 miles per hour to feet per second.

$$\left(\frac{63.9\,mi}{1\,hr}\right)\left(\frac{5280\,ft}{1\,mi}\right)\left(\frac{1\,hr}{3600\,\sec}\right) = \frac{93.72\,ft}{\sec}$$

The answer is 93.72 ft per second.
The units of measurement for miles and hours cancel.
The remaining measurements of ft in the numerator, and sec in the denominator result in the answer of feet per second.

(5)
A sphere has a radius of 2.3 yards.
A cylinder has a radius of 4 ft, and a height of 8 ft.
Which solid has the largest volume?

Volume of a Sphere:
$$V = \frac{4}{3}\pi r^3$$

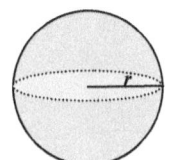

Volume of a Cylinder:
$$V = \pi r^2 h$$

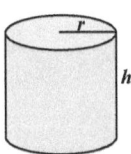

One solution plan is to convert the sphere radius to feet. Then, the sphere volume will be expressed as cubic feet, and the cylinder volume will also be expressed as cubic feet. The two volumes then can be compared to determine which solid has the largest volume.

Convert 2.3 yards to feet: $(2.3\,yd)\left(\frac{3\,ft}{yd}\right) = 6.9\,ft$

Volume of the sphere:
$$V = \left(\frac{4}{3}\right)(\pi)(6.9\,ft)^3 = 1376\,ft^3 \text{ (nearest whole number)}$$

Volume of the cylinder:
$$V = (\pi)(4\,ft)^2(8\,ft) = 402\,ft^3 \text{ (nearest whole number)}$$

Therefore, the sphere has the largest volume.

(6)
Equations illustrating Dimensional Analysis:

Determine which of the following equations are dimensionally correct.

(A)

$V = (s)(t)$, where $V = \dfrac{ft}{\sec}$, $s = ft$, and $t = \sec$

(B)

$t = \dfrac{V}{s}$, where $t = \sec$, $V = \dfrac{ft}{\sec}$, and $s = ft$

(C)

$s = \dfrac{at^2}{2}$, where $s = mi$, $a = \dfrac{mi}{\sec^2}$, and $t = \sec$

(D)

$s = (V)(t)$, where $s = ft$, $V = \dfrac{ft}{\sec}$, and $t = \sec$

Answers:

(A)
$V = (s)(t)$

$\dfrac{ft}{\sec} \neq (ft)(\sec)$

This equation is <u>not</u> dimensionally correct.
Therefore, the calculation would <u>not</u> be correct.

(B)
$t = \dfrac{V}{s}$

$\sec = \dfrac{ft/\sec}{ft} = \left(\dfrac{ft}{\sec}\right)\left(\dfrac{1}{ft}\right) = \dfrac{1}{\sec}$

$\sec \neq \dfrac{1}{\sec}$

The equation is <u>not</u> dimensionally correct.
Therefore, the calculation would <u>not</u> be correct.

(C)

$$s = \frac{(a)(t^2)}{2}, \text{ where } s = ft, a = \frac{ft}{\sec^2}, \text{ and } t = \sec$$

$$ft = \frac{\left(\frac{ft}{\cancel{\sec^2}}\right)(\cancel{\sec^2})}{2} = \frac{ft}{2}$$

The equation is dimensionally correct.
Therefore, the calculation can be correct.

(D)

$$s = (V)(t)$$

$$ft = \left(\frac{ft}{\cancel{\sec}}\right)(\cancel{\sec}) = ft$$

The equation is dimensionally correct.
Therefore, the calculation can be correct.

(7)
Convert 120 kilometers per hour to meters per second.

Solution:
(A)

$$\left(\frac{120\,\cancel{km}}{\cancel{hr}}\right)\left(\frac{1000\,m}{1\,\cancel{km}}\right)\left(\frac{1\,\cancel{hr}}{60\,\cancel{min}}\right)\left(\frac{1\,\cancel{min}}{60\,\sec}\right) = \frac{m}{\sec}$$

Equation (A) illustrates the result for the <u>dimensional analysis</u> only. The numerical calculations were not intended for this equation.

(B)

$$\left(\frac{120\,\cancel{km}}{\cancel{hr}}\right)\left(\frac{1000\,m}{1\,\cancel{km}}\right)\left(\frac{1\,\cancel{hr}}{60\,\cancel{min}}\right)\left(\frac{1\,\cancel{min}}{60\,\sec}\right) = \frac{33\frac{1}{3}\,m}{\sec}$$

Equation (B) illustrates both the <u>numerical</u> and the <u>dimensional</u> results.

Note that the cancelled factors are km, hr, and min, leaving m in the numerator and sec in the denominator for the desired result of meters per second.

(8)
Astronomical distances are sometimes described in terms of light-years. A light-year is the distance that light will travel in one year.

How far, in meters, and in miles, does light travel in one year? When the speed of light is given in meters per second, it is necessary to know how many seconds are in a year. This can be accomplished by converting units of measurement.

The speed of light is 299,792,458 meters per second.
Also, 1 year = 365.25 days, 1 day = 24 hours, 1 hour = 60 minutes, and 1 minute = 60 seconds.
Combining this information, the number of seconds in a year is:

$$\left(\frac{365.25 \text{ days}}{\text{yr}}\right)\left(\frac{24 \text{ hr}}{1 \text{ day}}\right)\left(\frac{60 \text{ min}}{1 \text{ hr}}\right)\left(\frac{60 \text{ sec}}{1 \text{ min}}\right) = \frac{31{,}557{,}600 \text{ sec}}{\text{yr}}$$

The distance, in meters, that light travels in one year:

$$1 \text{ light-year} = \left(\frac{299{,}792{,}458 \text{ m}}{1 \text{ sec}}\right)\left(\frac{31{,}557{,}600 \text{ sec}}{1 \text{ yr}}\right)(1 \text{ yr})$$

$$= 9.460730473 \times 10^{15} \text{ m}$$

$$= 9{,}460{,}730{,}473{,}000{,}000 \text{ m}$$

Convert the speed of light from meters per second to miles per hour:

$$\left(\frac{299{,}792{,}458 \text{ m}}{1 \text{ sec}}\right)\left(\frac{39.37 \text{ in}}{1 \text{ m}}\right)\left(\frac{1 \text{ mi}}{63{,}360 \text{ in}}\right) = \frac{186{,}282 \text{ mi}}{\text{sec}}$$

The distance, in miles, that light travels in one year is:

$$1 \text{ light-year} = \left(\frac{186{,}282 \text{ mi}}{\text{sec}}\right)\left(\frac{31{,}557{,}600 \text{ sec}}{\text{yr}}\right)(1 \text{ yr})$$

$$= 5.878612843 \times 10^{12} \text{ mi}$$

$$= 5{,}878{,}612{,}843{,}000 \text{ mi}$$

(9)
Convert a 24-hour day to seconds.

$$\left(\frac{24 \text{ hr}}{\text{day}}\right)\left(\frac{60 \text{ min}}{1 \text{ hr}}\right)\left(\frac{60 \text{ sec}}{1 \text{ min}}\right) = \frac{86{,}400 \text{ sec}}{\text{day}}$$

There are 86,400 seconds in one 24-hour day.

(10)
The Pythagorean Challenge

A man bought a 5-foot fishing rod, and then boarded a bus to go home. The bus driver told him that anything longer than 4 feet was not allowed on the bus. So, the man went back to the store, and returned to board another bus with the same rod. This time, the driver observed his purchase, but had no comment or objection. WHY?

[Reference the Appendix for the answer to the Pythagorean Challenge].

Section 4

This section contains more examples of complete dimensional analysis problems and solutions. The only noticeable difference from the previous sections may be detail and complexity.

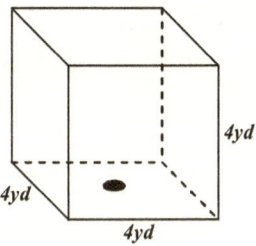

$$(4\,yd)(4\,yd)(4\,yd) = 64\,yd^3$$

(1)
A cubical water tank has an edge equal to 4 yards, and is filled completely with water. In the middle of the bottom side of the tank is a drain plug. When the drain is opened, the water will drain at the rate of 1 gallon per second.

Question:
What time is required to completely empty the tank?
Express your answer to the nearest second.

Given:

There are $\dfrac{7.48052\,gal}{ft^3}$

The drain rate is $1\,gal/sec$

Solution:

$$64\,yd^3 = (64\,yd^3)\left(\dfrac{27\,ft^3}{yd^3}\right) = 1728\,ft^3$$

$$(1728\,ft^3)\left(\dfrac{7.48052\,gal}{ft^3}\right) = 12926.34\,\text{gal in tank}$$

12926.34 gal = 12926 seconds (nearest second)

$$(12926\,sec)\left(\dfrac{1\,min}{60\,sec}\right)\left(\dfrac{1\,hr}{60\,min}\right) = 3.59\,hr \text{ (nearest hundredth)}$$

$(.59)(60\,min) = 35.4\,min$
$(.4)(60\,sec) = 24\,sec$

Therefore, the time required for the tank to empty is:
3 hours, 35 minutes, 24 seconds. ☺

(2)
Suppose someone told you that you would be given a million dollars in one dollar bills if you would provide a suitcase or some type of container to carry the money away.
Could this be accomplished?

Solution:
Each dollar bill weighs about .04 of an ounce.
Therefore, one million dollars will weigh:
$$(1,000,000)(.04\, ounce) = 40,000\, ounces$$

$$\left(\frac{40,000\, oz}{\frac{16\, oz}{1\, lb}}\right) = (40,000\, oz)\left(\frac{1\, lb}{16\, oz}\right) = \frac{40,000\, lb}{16} = 2,500\, lbs$$

Therefore, one million dollar bills weigh more than a ton, or about 2,500 pounds.

The answer to the question is NO.
Without human or mechanical assistance, one would not be capable of carrying the million dollars.

(3)
Suppose you had a billion one-dollar bills. The length of a dollar bill is 6.2 inches. Imagine one billion dollars aligned end-to-end.
Question:
What distance would be required for these one billion dollars?

Solution:
One billion can be expressed as 10^9.
$$(10^9)(6.2\,in) = 6.2 \times 10^9\,in$$

Converting the inches to miles:
$$(6.2 \times 10^9\,in)\left(\frac{1\,ft}{12\,in}\right)\left(\frac{1\,mi}{5280\,ft}\right) = 97{,}853.53535\,mi$$

The distance around the world at the equator is approximately 25,000 miles:
$$\frac{97{,}853.53535\,mi}{25{,}000\,mi} = 3.9141414 \text{ times around the world}$$

So, a billion dollar bills will extend almost 4 times around the world at the equator.

(4)
Chess – A Plethora of Possibilities

The first move (white and black) in a game of chess can be made in exactly 400 different ways. The possibilities for the first ten moves is more than 169 octillion, which is:
169,000,000,000,000,000,000,000,000,000.

A number followed by 27 zeros is really incomprehensible – so attempting to consider the total number of possible moves in any given lengthy chess game does overwhelm the imagination. But, back to my idea regarding the first ten moves that provide more than 169×10^{27} possibilities. How long would it take to play all possibilities for the first ten moves for each player? At the end of the year 2017, the earth's population was approximately 7.6 billion people. On the basis of this population, and 169 octillion possibilities, more than 42 million years would be required to play all the possibilities, even if every man, woman, and child in the world played without cessation for that duration at the rate of one game per **minute** with no game repeated.

$$\frac{169 \times 10^{27} \, games}{\left(7.6 \times 10^9 \, people\right)\left(60 \, \frac{games}{hr}\right)\left(24 \, \frac{hr}{day}\right)\left(365.25 \, \frac{days}{yr}\right)}$$

= 42,278,580,320,000 years for 7.6×10^9 people
(This is more than 42 **trillion** years).

WOW!

Now, to even consider the total number of possible moves in a lengthy chess game is, again, incomprehensible.

(5)
Question:
What is the capacity (volume) of the New City of Heaven (the New Jerusalem)?

When the Lord Jesus returns to our earth, He will be coming in judgment of all the people who have not trusted and received Him as Savior, and He will be coming to gather all of His Believers to reside with Him in His Heavenly Kingdom. This event is known as the Second Coming or Second Advent. There will be a new Heaven and a new Earth. The new Heaven, referred to as the New Jerusalem, will be the Holy City, having the Glory of God.

The dimensions of this Heavenly City are given in Revelation, Chapter 21 of God's Word, the Bible. The length, width, and height of the City are equal. Each of these dimensions is equal to twelve thousand furlongs.

Another way of describing the City is to define it as a cubical structure, each edge of which has a length of 12-thousand furlongs.

Answer:

A furlong is equal to $\frac{1}{8}$ of a mile.

$$(12{,}000 \text{ furlongs}) \left(\frac{\frac{1}{8} mi}{furlong} \right) = (12{,}000)\left(\frac{1}{8} mi\right) = 1{,}500 \, mi$$

So, each of the dimensions of the City of Heaven is equal to 1500 miles.

The Heavenly Cube

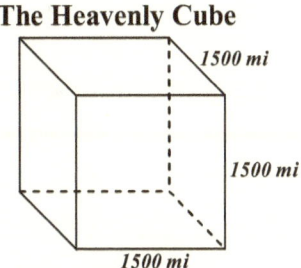

Therefore, the Volume of the Heavenly City =
$(l)(w)(h) = (1500mi)(1500mi)(1500mi) = 3{,}375{,}000{,}000 \text{ cubic miles}$

(6)
Question:
How long would it take to count to one million, counting at the rate of one number per second?

Solution:
Counting at the rate of one number per second will require 1 million seconds to count to one million.

$$\frac{1,000,000 \text{ sec}}{\frac{86,400 \text{ sec}}{\text{day}}}$$

$$(1,000,000 \text{ sec})\left(\frac{1 \text{ day}}{86,400 \text{ sec}}\right) = 11.57 \text{ days}$$

$$(.57 \text{ day})\left(\frac{86,400 \text{ sec}}{\text{day}}\right) = 49,248 \text{ sec}$$

$$\frac{49,248 \text{ sec}}{\frac{3600 \text{ sec}}{\text{hr}}} = 13.68 \text{ hr}$$

$$(.68 \text{ hr})\left(\frac{60 \text{ min}}{\text{hr}}\right) = 40.8 \text{ min}$$

$$(.8 \text{ min})\left(\frac{60 \text{ sec}}{\text{min}}\right) = 48 \text{ sec}$$

So, 11.57 days = 11 days, 13 hrs, 40 min, 48 sec

A Million may be <u>more</u> than you imagine!

(7)
There are two painters.
Painter #1 can paint a garage in one day.
Painter #2 can paint a garage in two days.
Working together, what time will be required for them to paint two garages?

Let g = one garage.
Let d = days that each painter works independently.
Let D = days that the two painters work together.

$\dfrac{g}{d}$ = one garage per day (Painter #1)

$\dfrac{g}{2d}$ = one garage per two days (Painter #2)

(1) $D\left(\dfrac{g}{d} + \dfrac{g}{2d}\right) = 2g$

(2) $D\left(\dfrac{2g+g}{2d}\right) = 2g$

(3) $D\left(\dfrac{3g}{2d}\right) = 2g$

(4) $D = (2\cancel{g})\left(\dfrac{2d}{3\cancel{g}}\right)$

(5) $D = \dfrac{4}{3}d = 1\dfrac{1}{3}$ days required for the two painters to paint two garages.

(8)
Assuming that fractional hens are as proportionally functional as whole hens, consider the following situation:

Question:
If a hen-and-a-half lay an egg-and-a-half in a day-and-a-half, how many eggs will three hens lay in three-and-a-half days?

To avoid possible confusion, the same question will be expressed differently:
If 1.5 hens lay 1.5 eggs in 1.5 days, how many eggs will 3 hens lay in 3.5 days?

	Hens	Eggs	Days
(1)	$3/2$	$3/2$	$3/2$

Line (1) above illustrates the initial status of this situation.

Solution:
Twice as many hens will lay twice as many eggs in the same number of days.

	Hens	Eggs	Days
(1)	$3/2$	$3/2$	$3/2$
(2)	3	3	$3/2$

In line (2), both the number of hens and the number of eggs have been multiplied by 2, so that now twice the number of hens have laid twice the number of eggs in the <u>same</u> amount of time.

We now have 3 hens laying 3 eggs in $3/2$ days. However, we want to know how many eggs 3 hens will lay in $3\frac{1}{2}$, or $7/2$ days.

If we multiply $3/2$ days by $7/3$, the result is $7/2$, or $3\ 1/2$. However, if the $3/2$ days is multiplied by $7/3$, the number of eggs must also be multiplied by $7/3$ to yield 7 eggs. So, line (3) indicates that 3 hens will lay 7 eggs in $7/2$ days.

	Hens	Eggs	Days
(1)	$3/2$	$3/2$	$3/2$
(2)	3	3	$3/2$
(3)	3	[7]	$7/2$

(9)
The current World Record for the 100 meter (m) sprint is 9.58 seconds by Usain Bolt. He established this record in 2009.

Given:
There are exactly 3,937 inches in 100 m.
There are approximately 1,609.347 m in one mile.

Using conversion factors and dimensional analysis, determine the solutions to the following two questions:

(1)
Running at his 100 meter rate, what would Bolt's time be for 100 yards?
[Express your answer to the nearest hundredth of a second].

(2)
Running at his 100 meter rate, what is Bolt's time for 100 meters in miles per hour?
[Express the answer to the nearest hundredth in miles per hour].

Solution for (1):
$$\left(\frac{9.58\,sec}{100\,m}\right)\left(\frac{100\,m}{3937\,in}\right)\left(\frac{36\,in}{1\,yd}\right)(100\,yd) = 8.76\,sec$$
So, Bolt's time for 100 yards would be 8.76 seconds.

Solution for (2):
$$\left(\frac{1\,m}{.0958\,sec}\right)\left(\frac{3600\,sec}{hr}\right)\left(\frac{1\,mi}{1609.347\,m}\right) = \frac{23.35\,mi}{hr}$$
So, Bolt's time for 100 meters is 23.35 miles per hour.

Note:
100 m = exactly 3937 in.
100 yd = exactly 3600 in.

The difference is 3937 − 3600 = 337 in.

$$\frac{337\,in}{12\,\frac{in}{ft}} = (337\,in)\left(\frac{1\,ft}{12\,in}\right) = 28.08\frac{1}{3}\,ft$$

So, 100 meters is $28.08\frac{1}{3}$ ft farther than 100 yards.

Note to Reader:
This is the same problem presented by problem (9) of Section 2. This alternate solution demonstrates a different analysis.

(10)
If a runner runs 100 yards in 10 seconds, what time is required (assume the same rate) to run 100 meters?
[Express the answer to the nearest hundredth of a second].

$$\frac{100 \text{ yd}}{10 \text{ sec}} = \frac{1 \text{ yd}}{? \text{ sec}}$$

$$(100 \text{ yd})(? \text{ sec}) = (1 \text{ yd})(10 \text{ sec})$$

$$? \text{ sec} = \frac{(10 \text{ sec})(1 \text{ yd})}{100 \text{ yd}} = .1 \text{ sec}$$

So, 1 yd = .1 sec

100 meters is exactly 337 inches further than 100 yards.

$$\frac{337 \text{ in}}{36 \frac{\text{in}}{\text{yd}}} = 9.36 \text{ yd}$$

$$(9.36 \text{ yd})\left(\frac{.1 \text{ sec}}{\text{yd}}\right) = .936 \text{ sec}$$

So, it requires .936 seconds to run an additional 337 inches. Therefore, running at the rate of 10 seconds for 100 yards, a runner would run 100 meters in 10 seconds + .936 seconds, or 10.94 seconds.

(11)
A Dimensional Analysis Challenge
    The Bondoraxpas

The sovereign nation of Gitvilan has developed a Doomsday Device that will discourage the aggressive intentions of other countries. Excluding any reference to the enormous destructive power source of this potent thermonuclear device, this weapon is comprised of the following significant components:

(1) Sisroters
(2) Treclecomereas
(3) Pyrogscoes
(4) Trevmegoritas

These four functionally significant components constitute the device that is known as the Bondoraxpas.

The relationship among the four components and the Bondoraxpas is as follows:

(A) There are 6 Sisroters for each Treclecomereas.
(B) There are 3 Treclecomereas for each Pyrogscoes.
(C) For each Pyrogscoes, there are 3 Trevmegoritas.
(D) There are 3 Trevmegoritas for the Bondoraxpas.

Using the four relationships above, develop the dimensional analysis equation that determines:

How many Sisroters are in the Bondoraxpas?

[Reference the Appendix for the Bondoraxpas Challenge].

Note to Reader:
This problem presents purely a mathematical challenge, without consideration of dimensional analysis and units of measurement.

(12)
(A) $3x - y = 12$

(B) $\dfrac{8^x}{2^y}$

Given the two related mathematical expressions (A) and (B), determine the solution.

[Reference the Appendix for the Mathematical Challenge].

Section 5

The information in this section has no relationship to the dimensional Analysis material presented in the previous four sections.

This section contains a miscellany of Thoughts and Ideas, some of which, may pique your interest. ☺

1

Some people have the idea that the concept of "Santa Claus" is only for children. There is a more significant perspective.

1

The redeeming quality of the fictional Santa Claus is that he is a symbol of the generosity and benevolence of our human spirit that, at least temporarily, suppresses the ugliness of our nature.

2

> Many people consider a funeral service as a sad, somber event. For many other people, some funerals are blessed occasions.

2

A funeral attendee is either a <u>Mourner</u> ☹, or a <u>Celebrant</u> ☺.
The Celebrant is trusting the Lord Jesus, and knows that the funeral service is for someone who also is trusting Jesus. Both will be residing with Jesus in His Heavenly Kingdom.

3

There are many "gods" in our world. They are the concoctions of those who do not trust the God of all Creation.

3

The Sovereign Lord God of all Reality never has an "oops" moment. This is just one characteristic that distinguishes Him from <u>all</u> the "little g" gods.

4

> Jesus is coming to this earth again.
> You need to be ready when He comes!

4

More certain than any event that has already occurred on earth is the absolute certainty that the Lord Jesus is coming again to our world.

He is coming in judgment for all who have rejected His Salvation, and He is coming to affirm His Salvation to all who have trusted Him.

Are you ready for His return?

5

You have many situations in life when you make decisions to choose, or reject something. There is only ONE decision that you must make that is the <u>most</u> <u>important</u> decision in your life.

5
Everyone has a Binary Choice of options in regard to God's offer of Salvation.

6

> Many people will reject genuine Faith.
> Genuine Faith is an Eternal Blessing to many other people.

6

Faith is having confidence in the Righteous Character of God & Jesus that will foster trust and assurance, even though our circumstances are fostering doubt and despair.

7

Are you intimately acquainted with the three G's?
This Trio is the most important and wonderful association that anyone could have.

7

The most significant and important gift that I could possibly receive was given to me many years ago – the three G's. I will have these three G's forever. Praise God & the Lord Jesus for God the **Father**, God the **Son**, and God the **Holy Spirit**!!!

8

God is the author of all Truth. Praise Him for His Love, Mercy, and Grace.

8
The Eternal Truth for all who will trust the Lord Jesus:
God will **Reign**, and the Son will **Shine**!

9

> There had to be a <u>first</u> Christian.
> Who is this Christian?

9
God is a **Christian**, and He is discriminating in His Judgment.

10

> Your time in this world is relatively brief.
> If you desire to have pleasant memories, carefully consider your thoughts, words, and actions.

10
My relatively brief remaining time on this earth is significantly decreasing the opportunities for creating memories.

11

Have you thoughtfully and honestly considered the nature and character of our national government? If you have, do you recognize the dysfunction, and corruption of many congressional members? They are selfish, and hypocritical. They lie and deceive. Tragically, they think we should be satisfied with their

11

I oppose the Plutocracy of wealth,
I cherish the Aristocracy of learning,
but I am grateful to the Sovereign Lord God & the Lord Jesus that some people of this world demonstrate the democracy of the human heart.

However, in general, the various governments of this world, <u>including</u> America, are being administered by kakistocracies! This deplorable situation is not unpredictable or unexpected, because of the sinful, unregenerate nature of many people that have not trusted and received the Salvation of the Lord Jesus Christ.

12

> Truth will ultimately prevail over everything that is false and wrong.
> God & Jesus will vanquish everything that is evil.

12
Only the Truth is always strong enough to overcome evil.

13

What is precious in your life? I know something that is precious to me. All goodness is a result of the Love, Mercy, and Grace of God and the Lord Jesus.

13
There are only a few things more precious than a loving, understanding heart.

14

> There is nothing in this world that is more important to me than the Peace of knowing that the Lord Jesus is my Personal Savior.

14
For Politics, Gold, and Silver, **I Do Not Lust**.
With Gratitude for the Love, Mercy, and Grace,
In the Salvation of the Lord Jesus, **I Do Trust**.

15

> Many people casually, carelessly, or incorrectly apply the concept of "miracle" to many events. There are genuine miracles, and you and I are included among them.

15
<u>All</u> Life is a Miracle (Created by God).
Life is <u>not</u> a Random Occurrence (Evolution).

16

> There <u>can</u> be a significant difference between Opinion and Fact. Sometimes the difference is inconsequential. There are times when your belief has <u>eternal</u> consequences.

16
You are entitled to your own OPINION, but you are not entitled to your own FACTS.

17

> Religious activities and involvement <u>cannot</u> and <u>will</u> not result in your Salvation. Only trusting Jesus will give you Eternal Life. Only Jesus Saves!!!

17

Regardless of your phony "religious" feelings, or even your genuine, sincere convictions, your religious activities will not reconcile you with God, or gain you access to His Heavenly Kingdom. Only Faith in Jesus <u>will</u>.

18

> Do you know that you will live <u>forever</u>? The only question that remains is <u>where</u> you will be forever?

18

A wonderful Truth is that I will live forever.
Even more wonderful, I will <u>not</u> be residing in this world, or Hell. I will be with Jesus.

Epilogue

In our Heavenly Kingdom residence, we (Christians) will enjoy a dimensionless existence. No longer will we be restricted and limited to the ubiquitous four dimensions of our present physical world, namely, the fundamental three dimensions, and time. In Heaven, dimensions will be irrelevant.
Praise the Sovereign God and the Lord Jesus Christ!!!

Postface

In this Postface, I digress from the subject of dimensional analysis to present a statement concerning the general climate of our nation's culture:

Appearance and reality are often diametrically opposed. In particular, I am referring to the deficient human qualities of Truth and Integrity. Regrettably, our corrupt and dysfunctional American Federal Government is analogous to a gold-plated watch, or a glass diamond ring. Such is our sinful human nature **without** the reconciliation of the Salvation, Love, Mercy, and Grace of the Lord Jesus Christ.

Appendix

Section 3, (10)
The Pythagorean Challenge
The man obtained a 3-foot by 4-foot box at the store, and placed the fishing rod inside the box diagonally.
The 5-foot fishing rod was positioned as the diagonal inside the box. Hence, the rod was the 5-foot hypotenuse of the 3-foot by 4-foot box, forming a 3,4,5 Pythagorean right triangle.

Section 4, (11)
The Bondoraxpas Challenge

$$\left(\frac{6\,Sisroters}{1\,Treclecomerers}\right) x \left(\frac{\cancel{3}\,Treclecomerers}{1\,Pyrogscoes}\right) x \left(\frac{1\,Pyrogscoes}{\cancel{3}\,Treymegoritas}\right)$$

$$x \left(\frac{3\,Treymegoritas}{1\,Bondoraxpas}\right) = \frac{18\,Sisroters}{1\,Bondoraxpas}$$

$$\frac{6 x \cancel{3} x 1 x 3}{1 x 1 x \cancel{3} x 1} = \frac{18\,Sisroters}{Bondoraxpas}$$

So, there are 18 Sisroters for each Bondoraxpas.

Section 4, (12)
The Mathematical Challenge
In (B), $\dfrac{\left(2^3\right)^x}{2^y} = \dfrac{2^{3x}}{2^y} = 2^{3x-y}$

Given in (A) that 3x-y = 12,
we have $2^{3x-y} = 2^{12}$
Therefore, the solution to (A) and (B) is $2^{12} = 4096$

www.ingramcontent.com/pod-product-compliance
Lightning Source LLC
Chambersburg PA
CBHW020441220526
45464CB00002B/798